North American
Animals of the Arctic

By Colleayn O. Mastin · *Illustrated by Jan Sovak*

Caribou

Both caribou and reindeer
Have two distinctive features:
One, they live in groups called herds
And two, they're restless creatures.

In spring they migrate northward,
To reach their "calving ground";
In fall they head for forests,
Where winter food is found.

In the spring thousands upon thousands of migrating caribou run swiftly across the vast tundra to reach the place where the baby calves will soon be born. The females lead the march while the bulls follow behind at a slower pace. Often the females will push forward, marching about twenty-four to thirty-two kilometers (fifteen to twenty miles) per day.

On the journey the caribou must swim across lakes and broad rivers. They are strong swimmers. The hollow hairs of their coats help them afloat, almost as if they were wearing a life jacket.

Just after the cow gives birth to a single calf, she gently nudges it to get up. It is a dangerous time for the calf, which must quickly become strong enough to run with the herd. If it doesn't it could be eaten by wolves, bears or foxes. Almost all the calves are born within a five-day period. This allows the herd to stay together and helps to increase the number of babies that survive.

Caribou eat grasses, arctic willow, moss and lichen.

The caribou is most familiar to northern peoples, for whom it provides both food and clothing. At one time native peoples made weapons, tools and toys from the antlers. Both sexes of caribou have antlers.

In Norway caribou are called reindeer.

Wolverine

A wolverine is fierce and strong
And very hard to scare;
Though of the weasel family,
It looks more like a bear.

A wolverine is darkish brown,
With a tan stripe down each side;
This cranky Arctic animal,
Would rather fight than hide.

For its size the wolverine is among the most powerful mammals in the Arctic. It has few enemies other than man.

A wolverine can drive away a cougar or even a bear from something one of these animals has killed. A wolverine is not large, but it has great strength. It is about the size of a bear cub. It is the largest of all the animals with musk glands, a group that includes the skunk and the weasel.

The wolverine is both a hunter and a scavenger. It will eat anything it can kill or find, including moose, beavers, birds, roots, or carcasses of caribou, seals and even whales.

Wolverines are not very social and hunt and live alone, except when raising their young. The mother gives birth to one to five babies. Her nest, which is usually hidden among the rocks, is lined with pieces of fur so it is warm and cozy for the young, which are called kits.

The kits are cared for by the mother, and they stay with her for about two years. Before she sends them away, she teaches them to hunt and to protect themselves.

Wolverine fur is used by northern people for trimming the hoods of their parkas. This fur has the ability to shed ice crystals, so it helps keep the frost from freezing their faces.

Dall Sheep

The dall is large, black or white,
With massive horns on its head;
In the mountains it feels at home,
On high ledges it hollows a bed.

It likes a wide view of the valley,
An enemy might be there,
For the dall doesn't want to be eaten
By a wolf or a cougar or bear.

Dall sheep live in herds of about ten animals. Usually, the oldest female, called a ewe, is the leader. In summer the females and males separate and only in the fall do the males rejoin the females and lambs.

Both female and male sheep have horns that continue to grow throughout their lives. The male sheep are famous for head-bashing contests that establish the strongest male in the herd. Two rams will run at each other, and bang their heads together so hard that the crash can be heard a kilometer (half a mile) away.

The chief enemy of the single baby lamb that is born to the ewe is the golden eagle. Other predators such as the wolf, wolverine and man are the main enemies of the adults.

Dall sheep eat a wide variety of plants and survive the winter by eating lichens. Their coats are white in northern regions, which makes it more difficult for them to be seen.

These sheep do not live in the high Arctic, but do live in the harsh, rocky areas of Alaska and northwestern Canada.

Arctic Fox

The Arctic fox just loves cold days,
And hunts the winter through;
Its summer coat is brown or gray,
In winter, white or blue.

For cozy toes on icy days,
It has four furry feet;
Lemmings, birds and "leftovers"
Are what it likes to eat.

The "leftovers" enjoyed by an Arctic fox are mostly provided by polar bears who have moved on after eating their fill of a seal, Arctic hare or caribou. This smart, cautious fox will follow Inuit hunters across the frozen land—at a safe distance, of course—to eat whatever they, too, may leave behind. The Arctic fox also feeds on dead animals, lemmings, mice, hares, birds and berries.

Some foxes are like squirrels and will store food away in dens in preparation for the long cold winter, when a supply of food is always in doubt.

The Arctic fox does not hibernate. Throughout the winter it makes its home in a burrow dug into the side of a hill or cliff. In spring anywhere from four to eleven pups are born. They are carefully looked after by both parents.

Like other Arctic mammals, the fox has both guard hairs and wool to keep it warm. The bottom of its feet are covered with long hair during the winter, which allows it to move quickly over the ice and snow.

After the snow has melted, the Arctic fox's snow-white coat turns light gray. This acts as a camouflage, so it is not so easily spotted by its enemies, which are mainly wolves and man.

The bark of the Arctic fox sounds something like the yapping of a noisy dog.

Arctic Hare

The Arctic hare is proud of its coat,
And it likes to keep it clean;
So it takes a bath in a snowbank,
Though the winter wind is mean.

Its favorite food is the willow tree,
Which it digs out of the snow;
This hare, in turn, makes tasty food,
As all Arctic people know.

In the high Arctic hundreds of these white hares gather on the tundra and stay together for warmth and protection from their many enemies. When a wolf or fox approaches, the hares hop away in different directions. This can be very frightening and confusing for their enemies.

These crafty hares also have been seen escaping enemies by heading toward a herd of muskox or staying close to humans until the hungry fox or wolf gives up and leaves. Then the Arctic hares quickly dash away, safe for a while.

Arctic hares have large furry feet that allow them to run easily across unpacked snow. To survive the cold they have relatively small ears, which help to preserve body heat. Often hares will tuck their tails between their hind legs, place their front paws in front of themselves and roll almost into a ball to keep warm.

For food the hares eat grasses, leaves and flowers, but during the winter their main food is the Arctic willow.

In the spring, between four and seven baby hares, called leverets, are born. At birth their eyes are wide open, and they can run within a few minutes.

The baby hares have thick, gray-flecked fur that blends perfectly into the tundra. This makes it very difficult to see them until the mother calls, when they scurry to her to nurse. This camouflage is another survival method.

Arctic Ground Squirrel

Chattering Arctic ground squirrels,
When its cloudy stay at home;
But when summer days are sunny,
They really like to roam.

They spend each long day hunting
For seeds and grassy roots;
They like to sit and nibble on,
Green plants and juicy fruits.

These curious and noisy Arctic squirrels are the only Arctic animals to hibernate for the entire long cold winter. To sleep through the eight months of winter, these little squirrels must dig burrows and line them with leaves and muskox hair.

Then they roll up into small balls with their bushy tails over their heads and nestle cozily in their dens to go to sleep. They wake up every few weeks to clean their fur and dens. Their homes are a maze of tunnels and burrows, which are also used to escape from their many enemies.

To watch for an approaching enemy, ground squirrels stand on their hind legs, carefully looking for any sign of movement. Should they see one of their many enemies—which include foxes, owls, ermine and hawks—they quickly dive into their dens for safety.

In the early summer four to eight young are born to the mother squirrel.

In the late summer the squirrels carefully store a supply of their favorite foods so they will have something to munch on when they wake and find snow still on the ground.

Arctic Wolf

The hungry wolf's a carnivore,
It eats both moose and mice;
In many children's storybooks
The wolf seems bad, not nice.

The wolves all travel in a pack
That isn't large in size;
Their coats are gray or brown or white,
But they all have yellow eyes.

The howling of a wolf pack is one of the scariest sounds in the wilderness. It is the way wolf packs talk to one another and may be their way of telling other packs to stay away from their territory.

Arctic wolves live in the northern parts of North America. For food they hunt elk, bison, mountain goat and sometimes, domestic or ranch animals. When food is scarce they hunt lemmings, hares and other small animals.

A wolf pack usually has eight to ten members. The leader can be either a male or female wolf. Wolves have no enemies other than man.

After a female wolf gives birth to a litter of five to fifteen hungry squealing pups, she stay in her den looking after them. Other members of the pack protect her and bring her food. Usually, only one female per pack gives birth to a litter.

Muskox

In the high frozen Arctic,
Where people seldom go,
Herds of muskox spend their days
Wandering through the snow.

They feed on tasty willow trees,
And other tundra plants;
These great herds are mostly made
Of mothers, kids and aunts.

Muskox live on the tundra of the far cold North and like to be together in large groups called herds. The herd is led by one or more of the strongest males. Muskox eat Arctic willow, lichen, grass and herbs.

Muskox cows usually have one calf every two years, so the herds grow very slowly. Both the males and females have huge horns.

Much like a wagon train under attack in a western movie, muskox make a circle facing outward when attacked by their main enemy, the Arctic wolf. Calves are placed within the circle for safety. Sometimes a bull or cow will charge out of the circle and try to gore the wolf with its sharp horns.

Of the over three thousand mammals that live in the world, only about forty of them are strong enough to survive in the Arctic. Of those, nearly one half live in the sea. No reptiles live in the Arctic and only about one hundred birds live there. Only ten species of birds stay all year round. The balance migrate south each winter.

Polar Bear

Polar bears live near the sea,
They're active all year round,
Stalking the frozen seascape
Wherever seals are found.

Polar bears can run and swim,
They're masters of survival;
Except for men, in their white world,
They have no Arctic rival.

Polar bears are very fierce! They live all over the cold Arctic region. These great white bears may be seen on ice floes, on bare rocky shores or on islands. They swim very well, holding their necks and heads out of the water.

Some of the polar bear's favorite foods are seal, stranded whales or other dead animals found on the shore. By far, the seal is the main prey. When hunting, the polar bear will sometimes cover its black nose. This makes it less noticeable to the creatures that it hunts for food.

In the winter the female bear digs a den in the snow and stays there to have her cubs. The one or two cubs born in this winter den are so small at birth that you could hold one in your hand.

Until they are almost two years old, the mother bear protects her cubs and teaches them how to survive. Then she sends them away to have families of their own.

Harp Seal

The furry harp seal gets its name
From the harp-shaped marks on its back;
Though it spends its life in the ocean,
It is born on the cold ice pack.

These seal pups don't get babied;
In two weeks they're on their own,
For then their mothers abandon them,
And they enter the ocean alone.

Every February dense herds of female harp seals climb up onto the floating ice packs to bear their young.

The harp seal pups must learn very quickly to fend for themselves. After being fed on their mothers' rich, fatty milk for just two or three weeks, they have grown very quickly and now weigh ten times more than they did at birth.

Although they are not yet fully grown, some of the pups are strong enough to survive when their mothers leave them. In the ocean they can dive long and deep to catch the fish that is their food.

When the pups are born their fur is entirely white, but as they grow older their coats turn gray and their faces turn black.

On the ice they can flipper themselves along quickly if they have to escape from an enemy. Their many enemies include polar bears, whales and man.

Narwhal

Though it looks like a fish, it's a narwhal,
A mammal at home in the sea;
Up in the cold Arctic Ocean
Is where it prefers to be.

It's called a "Sea-Going Unicorn"
Because of its long, long tooth;
No one really knows why it has one,
If you want to know the truth.

A narwhal's long spiraling ivory tooth may grow to be three meters (ten feet) in length.

Usually, the male has a tusk, but sometimes a female will have one, too.

Long ago, many people believed these tusks had magical powers. Narwhals are hunted by northern peoples for food and for its ivory tusk that carvers treasure.

Narwhals are very social creatures and like to stay together in groups called pods. A pod usually consists of from five to twenty individuals.

In the cold icy waters where they live, they feed on fish, octopus, squid and crustaceans. Their enemies are polar bears, other whales and man.

Every two years female narwhals give birth to a single calf, which they nurse for up to two years. Narwhals, like other Arctic whales, do not have a top fin (called a dorsal fin) because it would get in the way when they are swimming under the ice.

Beluga Whale

A beluga whale is noisy;
It can sing and whistle and scream;
Its shortish body is round and plump,
With skin the color of cream.

This whale is not a great diver;
It likes shallow waters instead;
When it wants to get a breath of air,
It breaks through the ice with its head.

As soon as a mother beluga has her baby, which is called a calf, she pushes it to the surface for its first breath of air.

The baby calf must then nurse so it can quickly grow a thick layer of blubber that will keep it warm. At birth the baby is bluish gray in color and turns whiter as it grows older.

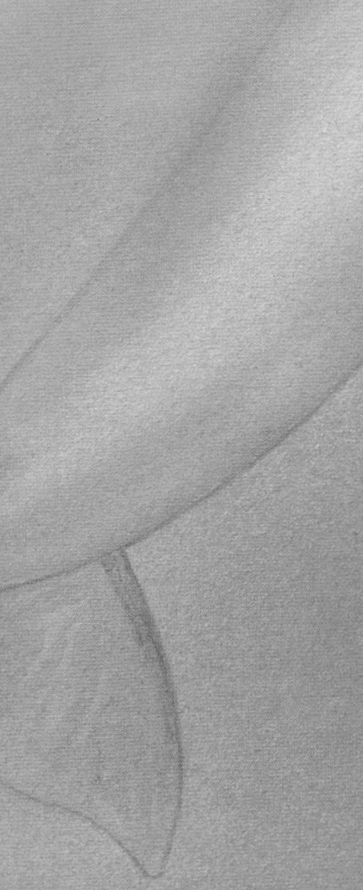

These expert swimmers stay close to the shores of the Arctic Ocean and some large rivers. They migrate each year, leaving the northern Arctic waters in the fall and returning in the spring. Belugas travel together in groups called pods of between ten and two hundred members.

Belugas eat mollusks and fish like suckers and catfish that live at the bottom.

Belugas love to sing. Whalers used to call them sea canaries because of the whistles, screams, moos and trills that make up their songs.

The orca, or killer whale, is the main enemy of the beluga. For survival it stays as far away as possible from this deadly hunter.

Walrus

A walrus spends sometime at sea;
Sometimes it stays on shore;
It lives in herds that number
Two thousand, maybe more.

A walrus weighs about a ton,
A creature of great power;
This mighty, graceful swimmer
Can dive deep for half an hour.

The only natural enemies of the walrus are the orca (the killer whale) and the polar bear. These are the only enemies with teeth strong enough to chew through the walrus's tough skin.

When a walrus herd smells a human hunter, the entire herd will quickly leave the land and dive into the sea. When one walrus in the herd is attacked, others will come to its defense. An injured walrus will be helped onto the sea ice or land by other members of the herd.

Every other year a walrus cow gives birth to one or two calves that stay with her for about three years.

Walruses use their tusks to break breathing holes in the ice and to haul themselves out of the water. Their tusks are also used to dig up clams, oysters, abalone and other mollusks on which they feed.

The males' long, sharp ivory tusks also are used to scare off other males.

Lemming

Lemmings are small and furry and brown,
And look something like a mouse;
They weave together bits of grass
To make themselves a house.

Lemmings do not hibernate.
They scurry all year round
Eating berries, buds and mosses
And roots found underground.

The chubby, furry brown lemming is a favorite food for many Arctic animals.

Foxes, wolverine, weasels, ground squirrels and wolves all hunt this little creature. Birds like the snowy owl will perch on a hill, watching for a lemming's slightest move, then swoop down for their dinner.

Lemmings are safer from their enemies in the winter, when they can hide under the snow. But they are never safe from foxes and ermine that hunt them all through the year.

When there is much food available, the female lemming gives birth to many tiny babies. Some years she gives birth to a litter of four to eight young every three to four weeks.

Every three or four years lemmings have a population explosion, which may cause overcrowding and a shortage of food. Because of this, these quarrelsome little creatures will scatter to find new places to live. Some will die trying to cross rivers, some move to areas where there is little food and starve, and others are eaten by their many enemies.

After this population explosion, the number of lemmings decreases again and the cycle of small populations, then large ones, begins again.

The Inuit call the lemming *kilangmiutak,* which means "one that comes from the sky." In Scandinavia, where lemmings also live, some people call them sky mice because in the years when there are so many of them, it seems as though they fall like rain from the sky.

Ermine

The graceful, cunning ermine
Has a black tip on its tail;
Its lovely fur is brown or white,
Its pointed face is pale.

Ermine sleep throughout the day,
At dusk they like to prowl;
When two lone ermine chance to meet
They bare their teeth and growl.

This fierce efficient hunter preys on other small animals and birds. It has a wonderful sense of smell and hunts for its food during the dark Arctic nights.

An ermine, or weasel as it is most commonly called, pounces on its prey, bites it at the back of the neck, then wraps its snakelike body around the victim so it cannot escape.

It hunt birds, lemming, rats, and eggs from birds nests. An ermine is an excellent swimmer, so it can easily catch small fish for dinner.

Ermines have a long sleek body and can squeeze through tiny openings to find food. They store food away for future use, especially during the long cold winter.

When an ermine is in danger it sprays a terrible odor that forces some of its enemies to leave. Its main enemies are wolves, foxes and bears.

When a female ermine is about to have babies, she makes a cozy nursery lined with fur from other animals. This nursery is in an underground tunnel usually among rocks or old buildings. The babies grow very rapidly, and they can have their own babies before they are one year old!

Published by
Grasshopper Books Publishing
106 Waddington Drive
Kamloops, British Columbia
Canada V2E 1M2

This book is dedicated to my favorite second oldest daughter, Michelle Colleen, and to Morris Locken.

Acknowledgments:
Solomon Awa, Nunavut Planning Commission, Grant Scott, Josh Lockwood, the governments of Canada and British Columbia, Encyclopedia of Mammals, *Emelee Marchant, Bill Gilroy, Wildlife Park, British Columbia.*

Canadian Cataloguing in Publication Data
Mastin, Colleayn, O. (Colleayn Olive)
North American Animals of the Arctic

(Grasshopper series; 2)
Includes index.
ISBN 1-895910-23-4

1. Zoology—Arctic regions—Juvenile literature. 2. Zoology—North America—Juvenile literature. I. Sovak, Jan 1953– II. Title. III Series: Mastin, Colleayn O. (Colleayn Olive), Grasshopper Series; 2. QL736.M385 1997 j591.9719 C97-910363-0

Printed in Canada

Index	***Page***

ECONOMICS IN THE MOVIES

G. DIRK MATEER

The Pennsylvania State University

Australia • Brazil • Japan • Korea • Mexico • Singapore • Spain • United Kingdom • United States

Economics in the Movies, 1e

G. Dirk Mateer

VP/Editorial Director:
Jack W. Calhoun

VP/Editor-in-Chief:
Dave Shaut

Acquisitions Editor:
Michael Worls

Sr. Developmental Editor:
Susan Smart

Marketing Manager:
John Carey

Production Editor:
Robert Dreas

Sr. Technology Project Editor:
Peggy Buskey

Web Coordinator:
Karen Schaffer

Manufacturing Coordinator:
Sandee Milewski

Production House:
OffCenter Concepts

Printer:
Malloy Lithographing Inc.
Ann Arbor, Michigan

Art Director:
Michelle Kunkler

Cover Designer:
Ann Small/a small design studio

Cover Images:
© The Image Bank

Printed in the United States of America
7 8 9 10 11 14 13 12 11

ISBN-13: 978-0-324-30261-5 package
ISBN-10: 0-324-30261-4 package

ISBN-13: 978-0-324-31676-6 book
ISBN-10: 0-324-31676-3 book

Library of Congress Control Number:
2004115969

For more information contact
South-Western Cengage Learning,
5191 Natorp Boulevard,
Mason, Ohio 45040.
Or you can visit our Internet site at:
academic.cengage.com

Contents

Preface

South-Western, a part of Cengage learning, proudly presents *Economics in the Movies.* This collection of twenty film scenes highlights many important economic principles, theories, and concepts. Specially designed as a student workbook to accompany any principles of economics textbook from South-Western Cengage Learning, it will help you learn core ideas through the magic and power of film.

This book is the product of my ongoing cinema-based teaching and research. I have found that students respond favorably to the linkage between film scenes and abstract theories or concepts. Moreover, film offers a visualization of economic concepts that are often hard to communicate through lectures or in a textbook. For example, one can read about the efficiency of the market using the traditional cost curve analysis. However, a visualization where Ben Stiller and Jennifer Aniston try to figure out the most effective way to run their lives in *Along Came Polly* (2004) can dramatically reinforce how well the concept of efficiency is learned.

A brief introduction provides a short overview of economics in film, music, and other media. This introduction will give you an appreciation for how this workbook was developed and some of the other innovative ways that are used to teach economics.

Instructions are provided in "Film Scenes for Learning Economic Principles, Theories, and Concepts." In this section you will find film and scene descriptions for twenty topics. Each of the film scenes is presented in a convenient two-page format. The first page describes the film and the scene that was selected; the second page has a space for your name, your analysis, and personal reactions.

A list of questions and issues to think about are provided while you view the scene. There is also a list of concepts or examples to motivate your reaction to the scene. Marking each concept off in the box provided will help you write your analysis.

You can use this book in different ways. Your instructor might show a scene in class for discussion. Or your instructor might assign scenes for your out-of-class viewing. He or she may also want you to write an analysis and turn it in for class credit.

This book also is useful for independent study. All of the film scenes are available to you on-line. Complete access instructions are provided on page 3. You can supplement and enhance your knowledge of economics by viewing the scenes that are linked to the topic you are trying to learn.

The film descriptions in this book were based on the latest versions of the following film reference sources:

- *Leonard Maltin's Movie & Video Guide*
- *Roger Ebert's Movie Yearbook*
- *TLA Video & DVD Guide*

Another valuable reference that was used is the Internet Movie Database (*www.imdb.com*). If you want to learn more about a specific film, I suggest that you use one of the reference books mentioned or the Internet Movie Database, where you can search for casting information, reviews, awards and nominations, bloopers, and more.

Many people helped to bring this project to a successful completion. I am especially grateful to David Shapiro at Penn State University for his encouragement in utilizing films in the classroom. I also want to thank Jim Gwartney and David MacPherson at Florida State University for mentoring the process and the vision they helped provide. In addition, I want to thank the thousands of students who have helped shape this work by providing film suggestions and feedback on what works best in the classroom. Without their enthusiasm and feedback, this project would not exist!

Months of dedicated work went into identifying scenes, editing, and licensing the talent that appears in the scenes. Each actor or actress signed an agreement so that we could bring theses scenes to you. Curtis Bowden, Dave Reeder, Cameron Dieterich, and Lisa Soboslai at Corbis Corporation handled this complex process superbly. Their enthusiasm and dedication in getting everything right was amazing!

I'd also like to thank everyone at South-Western Cengage Learning, for believing in the idea and helping launch this project—the only one of its kind in economics. Special thanks go to Susan Smart, Peggy Buskey, Bob Dreas, and John Carey for their efforts and encouragement.

Also, none of this would be possible without the support and love of my wife, Leslie, and children, Noelle and Nick. I am indebted to each of them for their patience and understanding!

An effort such as this workbook is always an ongoing process. Please send me feedback about any aspect of its content or design. You can send your comments to me directly at Department of Economics, 606 Kern Building, The Pennsylvania State University, University Park, PA 16802 USA. You can also send email to dmateer@psu.edu.

G. Dirk Mateer

Introduction to Economics in the Movies

This workbook is unique because it borrows from feature films in a way that enhances core content. Concepts are visualized by utilizing short film scenes. Others have utilized films to teach economics, most notably, Leet and Howser (2003), who developed a semester-long course that uses full-length films. However, in traditional 50- and 75-minute classes, full-length films are not a viable option nor can the instructor easily require students to see a film outside of class. Here you will find the best of Leet and Howser packaged into short film clips that are one to four minutes in length. Not surprisingly, there are similarities between what Leet and Howser recommend and the scenes chosen here. *It's a Wonderful Life* and *Erin Brockovich* are two films used by Leet and Howser that also find their way into this workbook. What distinguishes this effort from their work is that this book takes many films that appear to be unrelated to economics and utilizes short scenes to concisely capture key concepts. By drawing scenes from an eclectic mix of films this workbook shows how economic theory applies in areas that are not traditionally viewed as ripe for economic analysis.

For those of you who want to learn more about economics in film there are a number of useful lists for specific fields. Shor (2004) has compiled a comprehensive list of films that highlight game theoretic applications like the prisoner's dilemma, strategic behavior, and bargaining. Mulder (2002) catalogs how current labor issues are treated in the movies. Ribstein (2004) and Formaini (2001) explore how capitalism is portrayed in film.

Moreover, nontraditional approaches to teaching economics extend beyond film. Becker and Watts (1998, 2001, 2003) examined the way that economics was taught at the college level and found that the discipline had been slow to adopt innovative approaches to teaching. Some examples of non-traditional approaches include: Watts' (1998) inspection of economics in literature and drama, Kish-Goodling's (1998) application of *The Merchant of Venice* to teach concepts in money and banking, and Tinari and Khandke's (2000) use of popular songs, dating back to the 1930s, to help teach economics.

The point of this workbook (and the work of the others mentioned here) is to begin to see the world the way economists do. So long as scarcity exists and people make choices, economic decisions will play a key role in determining the actions of participants. My hope is that this workbook will help you take what you have learned and apply it in your everyday life.

Now on to the film scenes!

Film Scenes for Learning Economic Principles, Theories, and Concepts

This section has film descriptions, scene descriptions, discussion questions, and a list of concepts or examples for twenty economic topics. Each scene appears in a convenient two-page format that lets you write your scene analysis on the second page. You also have space to include your personal reactions to the scene.

The film description summarizes the film's plot to help you understand the scene in the context of the entire film. Scene descriptions set the context of the scene within the film by describing what occurs before, during, and after the scene. If you have not seen the film, or do not recall it easily, these descriptions will give you enough detail to understand the scene.

Some film scenes described in this book are either edited versions of the film's original scene or a composite of scenes from different parts of the film. If you have seen any of these films, do not be surprised if you cannot recall the exact scene presented to you. The scene descriptions tell you whether a scene was edited from the original or compiled from multiple scenes in the same film.

You can view the film scenes on the Internet by going to:

http://economicsinthemovies.swlearning.com

This URL takes you to the registration page for this film scene collection. You register once using the serial number that comes with this book. After registering on your first visit, you can return to view the scenes by logging in with the username and password you have selected.

Your instructor has the same film scenes on either a VHS tape or CD-ROM. She or he can show you the scenes during your regular class meeting or might ask you to view them as an outside assignment.

You can also view trailers for most of the films included in this book by going to:

http://www.imdb.com

Search for the film's title and then click on the link that takes you to the film that you are interested in. Along the left-hand column there is a link that reads "trailers." Watching the trailer can refresh your memory (if you have seen the film) or provide a useful overview of a film you have not seen.

The Family Man

Color, 2000
Clip Running Time: 2 minutes, 52 seconds
Rating: PG-13
Director: Brett Ratner
Distributor: Universal Studios Home Video

Jack Campbell (Nicolas Cage, 1996 Best Actor for *Leaving Las Vegas*) is a successful New York City investor. In *The Family Man* his life takes an unusual turn when he receives a phone message from his college sweetheart, Kate (Téa Leoni), whom he hasn't spoken with in years. Jack does not return the call, but when he wakes up the next morning, he finds himself in a parallel universe where he is married to Kate, works as a salesman for his father-in-law, and has a family. The movie asks us to imagine "what if" his life had been different and the two of them had married and lived an alternative life in the suburbs. Like Charles Dickens' *A Christmas Carol* or Frank Capra's *A Wonderful Life,* the notion of what could have been (opportunity cost) is a central theme.

Scene: A New Job

In the scene, Jack shows Kate an apartment that they could stay in indefinitely if they relocate to New York City. The apartment is luxurious and a big upgrade from their home in New Jersey. Jack tries to convince Kate that moving back into the city will improve their life by providing more rewarding jobs and better schools for their kids.

What to Watch for and Ask Yourself

- What are the trade-offs involved in moving?
- What is the opportunity cost of moving to the city? What is the opportunity cost of staying in New Jersey?
- Jack wants a "perfect life." Do you think that this is possible?
- Is the *ceteris paribus* assumption violated when Jack and Kate's lives are shown as a married couple? Explain why "holding everything else constant" is such a hard thing to do.
- Kate refutes Jack's point about living a life that others will envy by saying that "they already do envy us." Why do you think that is the case?

Concepts or Examples Please check the concepts you identify in the clip.

- ☐ Rationality assumption
- ☐ *Ceteris paribus*
- ☐ Opportunity cost
- ☐ Trade-offs
- ☐ Wants
- ☐ Self-interest
- ☐ Needs
- ☐ Subjective values

Analysis Write a short analysis using the concepts you checked.

____

____

____

____

____

____

Personal Reactions

____

____

____

____

____

____

Out of Sight

Color, 1998
Clip Running Time: 1 minute, 25 seconds
Rating: R
Director: Steven Soderbergh
Distributor: Universal Studios Home Video

Jack Foley (George Clooney) is a career bank robber who breaks out of prison, but he ends up locked in the trunk of a car with a federal marshal, Karen Sisco (Jennifer Lopez), when his escape does not end up as planned. The pair become attracted to one another, but it is her job to arrest him, and they pursue an improbable romance as the movie unfolds at a hectic pace.

Scene: The Price is Going Up

Richard Ripley (Albert Brooks) is a convict who makes the mistake of mentioning that he has a large stash hidden at his place when he gets out. So Maurice "Snoopy" Miller (Don Cheadle) and his associate Himey (James Black) shake him down. They extort Richard by charging him inflated prices to procure the items he wants.

What to Watch for and Ask Yourself

- Who seems to be getting the better end of the exchange?
- Why is Richard willing to pay Maurice's prices?
- This exchange is not entirely voluntary; Richard doesn't have other options like he would if he was out of prison. This gives Maurice a degree of monopoly power.
- If Maurice is seen as a monopolist, do the prices that are charged make more sense?
- Describe the consumer surplus of Richard and the producer surplus of Maurice.

Concepts or Examples Please check the concepts you identify in the clip.

- ☐ Exchange creates value
- ☐ Value
- ☐ Opportunity cost
- ☐ Trade-offs
- ☐ Wants
- ☐ Consumer surplus
- ☐ Needs
- ☐ Producer surplus
- ☐ Supply and demand
- ☐ Elasticity of price demand

Analysis Write a short analysis using the concepts you checked.

____

____

____

____

____

Personal Reactions

____

____

____

____

____

____

Seabiscuit

Color, 2003
Clip Running Time: 1 minutes, 30 seconds
Rating: PG-13
Director: Gary Ross
Distributor: DreamWorks and Universal Studios Home Video

Based on a true story, *Seabiscuit* it set against the backdrop of the Great Depression. The nation was looking for something to help escape the grind of daily life. As a result, Seabiscuit was the top news story of 1938, surpassing both the Depression and the rise of Hitler. The movie has a distinctly populist message: If an underdog like Seabiscuit could win against larger and better-bred horse like War Admiral, then maybe there was a chance for the average person to escape poverty and live a better life. The movie follows the lives of three men who are brought together: Seabiscuit's owner (Jeff Bridges), jockey (Tobey Maguire), and trainer (Chris Cooper). We see how each of them finds something to live for through the horse. Historian, David McCullough, narrates the film, which helps the viewer to appreciate the economic forces at work. *Seabiscuit* was nominated for seven Academy Awards including best picture.

Scene: The Stock Market Crash

The movie fades to a series of authentic black and white stills from the Depression era and then it fades back to color in a homeless camp. Billions of dollars of wealth have been lost and unemployment reaches 25 percent.

What to Watch for and Ask Yourself

- Using your textbook or the Internet, lookup unemployment rates over the last 30 years. In what year was the unemployment rate the highest?
- Compare the unemployment rate that you found with the rate of unemployment during the Great Depression.
- In October of 1987 the Stock Market crashed, losing 22 percent of its value on a single day. The size of the daily loss, on a percentage basis, was twice that of the crash of 1929. What happened to the economy immediately after the crash of 1987?
- Is the "crash" responsible for the Great Depression?
- What are some of the chief differences in monetary and fiscal policy that helped the economy stabilize in 1987 and avoid another depression?

Name ______________________________ **Seabiscuit**

Concepts or Examples Please check the concepts you identify in the clip.

- ☐ Great Depression
- ☐ Expectations
- ☐ Business Cycle
- ☐ Potential GDP
- ☐ Discouraged workers
- ☐ Trough
- ☐ Unemployment
- ☐ Contraction
- ☐ Actual GDP

Analysis Write a short analysis using the concepts you checked.

____

____

____

____

____

____

Personal Reactions

____

____

____

____

____

____

The Man Who Wasn't There

Black and White, 2001
Clip Running Time: 1 minute, 42 seconds
Rating: R
Director: Joel Coen
Distributor: Universal Studios Home Video

This Coen Brother's film (see also *Raising Arizona; Fargo; The Big Lebowski; O Brother, Where Art Thou?*) is set in Santa Rosa, California in the late 1940s. The movie was purposefully shot in black and white. This treatment makes *The Man Who Wasn't There* look like a classic film; however, the Coen Brother's have added a contemporary story line. Ed Crane (Billy Bob Thornton) is a barber who decides to blackmail the man who is having an affair with his wife.

Scene: Dry Cleaning

Ed Crane is cutting Creighton Tolliver's (Jon Polito) hair. Creighton Tolliver is in town to secure venture capital for a new enterprise in dry cleaning that he would like to start. Since Creighton Tolliver is a salesman, he launches into a long-winded discussion of his idea and the need for capital to make it happen. Pay particular attention to how Creighton describes the dry cleaning idea and how he intends to develop the business over time.

What to Watch for and Ask Yourself

- Why is capital a necessary ingredient for economic growth?
- What is venture capital and how is it different from ordinary capital?
- In the scene, what makes "dry cleaning" a risky venture?
- Why does cash flow allow a business to grow?
- Do you agree with Creighton Tolliver when he describes "dry cleaning as the biggest business opportunity since Henry Ford"?

Concepts or Examples Please check the concepts you identify in the clip.

- ☐ Productivity
- ☐ Innovation
- ☐ Capital
- ☐ Growth
- ☐ Advances in technology
- ☐ Investment
- ☐ Factors of production
- ☐ Production possibilities curve

Analysis Write a short analysis using the concepts you checked.

____

____

____

____

____

____

Personal Reactions

____

____

____

____

____

____

Waterworld

Color, 1995
Clip Running Time: 2 minutes, 5 seconds
Rating: P-13
Director: Kevin Reynolds
Distributor: Universal Studios Home Video

In the distant future, the greenhouse effect has melted the polar ice caps and the earth is covered by water. Civilization, as we know it, is lost under the sea. A mutant mariner (Kevin Costner), capable of swimming underwater, sails the oceans in his trimaran trying to survive. The oceans are populated by different groups of people, all of who seek dry land, something that no one can ever recall seeing. Eventually, the mariner ends up with two passengers: a woman, Helen (Jeanne Tripplehorn), and her adopted daughter, Enola (Tina Majorino), who has a map tattooed on her back that leads to dry land. However, before they can find the dry land, they must escape from a group of savage bandits intent upon seizing the map.

Scene: Trading Dirt

The mariner has in his possession a bag of dirt, something we'd take for granted today, but not in *Waterworld*. The dirt is more valuable than pure hydro (fresh water). Once the dirt is weighed and assessed for purity it is exchanged for 124 chits (or credits).

What to Watch for and Ask Yourself

- What makes dirt more valuable than hydro or chits?
- What makes dirt a good choice as money?
- How does the use of dirt improve the efficiency of the barter process and eliminate the need for a double coincidence of wants?
- Do the traders desire dirt for its asset demand or transactions demand?

Concepts or Examples Please check the concepts you identify in the clip.

- ☐ Medium of exchange
- ☐ Unit of account
- ☐ Store of value
- ☐ Barter
- ☐ Asset demand for money
- ☐ Relative scarcity
- ☐ Purchasing power
- ☐ Double coincidence of wants
- ☐ Transactions demand for money

Analysis **Write a short analysis using the concepts you checked.**

____

____

____

____

____

____

Personal Reactions

____

____

____

____

____

____

Reality Bites

Color, 1994
Clip Running Time: 1 minute, 11 seconds
Rating: PG-13
Director: Ben Stiller
Distributor: Universal Studios Home Video

Lelaina Pierce (Wynona Ryder) is a recent college graduate searching for a job and meaning in life. To chronicle life after graduation, Lelaina uses her camcorder to create a mock documentary of her friends and their post-graduation experiences. The promise of a fulfilling job seems like a remote possibility to these recent graduates who struggle to make ends meet and also find themselves. The result is a complex movie that explores relationships, materialism, and the meaning of life in a clever and amusing way. Ethan Hawke, Ben Stiller, and Renee Zellweger also star.

Scene: A Loan from Mom and Dad

Lelaina is valedictorian of her class, but she has trouble finding a job in her field. In this scene she is having a conversation with her mother (Swoosie Kurtz) and father-in-law about a loan. Her mother suggests that Lelaina should take an entry-level job because the local economy is struggling.

What to Watch for and Ask Yourself

- If you were Lelaina, would you want to work at a fast-food restaurant?
- What type of unemployment does this scene illustrate (frictional, structural, or cyclical)?
- Why might someone who was successful in college struggle to find and keep a job upon graduation?
- Why does Lelaina's father-in-law suggest to her that she omit her qualifications from her application?

Concepts or Examples Please check the concepts you identify in the clip.

- ☐ Discouraged workers
- ☐ GDP gap
- ☐ Potential GDP
- ☐ Frictional unemployment
- ☐ Production possibilities curve
- ☐ Actual GDP
- ☐ Structural unemployment
- ☐ Cyclical unemployment

Analysis Write a short analysis using the concepts you checked.

____

____

____

____

____

____

Personal Reactions

____

____

____

____

____

____

The Major and the Minor

Color, 1942
Clip Running Time: 3 minutes, 5 seconds
Rating: Not Rated
Director: Billy Wilder
Distributor: Universal Studios Home Video

Susan Applegate (Ginger Rogers) masquerades as a twelve-year-old to get a half-price ticket home. Part of the charm is that it is apparent to almost everybody, except Major Philip Kirby (Ray Milland), that she can't possibly be twelve! The trouble begins when Kirby's fiancé boards the train and she learns that Susan was allowed to sleep in the lower berth of the Major's compartment. So to prove that that Susan really is twelve, the Major enters her in school while she waiting to be taken home. A series of amusing misunderstandings ensue that make this screwball comedy memorable.

Scene: Getting the Half-Fare

After living in New York City for a year, Susan Applegate decides to return to Iowa. She has planned carefully and saved the *exact* train fare to get back home in a sealed envelope in case she would ever need it. However, her planned return does not allow for inflation. Prices have risen and she cannot afford to purchase a ticket. When she overhears that half-price tickets are offered for those younger than twelve, she comes up with an innovative scheme to purchase the discounted fare by soliciting the help of a man who is searching for coins left in telephones.

What to Watch for and Ask Yourself

- How does inflation impact real income?
- The train company raises prices from $27.50 to $32.50. What is the percentage change in nominal terms?
- How does inflation alter personal behavior?
- Susan eventually gets a half-price ticket. How effective is the company's price discrimination strategy? Comment by considering how likely it is that someone can misrepresent how old he or she is for financial gain.

Name ______________________________

Concepts or Examples Please check the concepts you identify in the clip.

- ☐ Real Income
- ☐ Nominal Income
- ☐ Cost of living adjustments (COLAs)
- ☐ Inflation
- ☐ Demand-pull
- ☐ Cost-push
- ☐ Price level
- ☐ Price discrimination

Analysis Write a short analysis using the concepts you checked.

____

____

____

____

____

Personal Reactions

____

____

____

____

____

It's a Wonderful Life

Black and White, 1946
Clip Running Time: 3 minutes, 39 seconds
Rating: Not Rated
Director: Frank Capra
Distributor: Republic Pictures Entertainment, A Paramount/Viacom Company

Considered one of the best films of all-time (AFI, 1998), *It's a Wonderful Life* has become an enduring holiday classic. George Bailey (James Stewart) dreams of leaving Bedford Falls are sidetracked for the good of those he loves. George sacrifices his ambitions to help restore the family business and prevent Mr. Potter from controlling the town. Along the way he helps the citizens of Bedford Falls to enjoy a better life. The only problem is that George Bailey can't see all the good he has done or the love of the people around him. He becomes despondent and tries to kill himself until he is rescued by his guardian angel, Clarence (Henry Travers), who tries to convince George that he has had a "wonderful life."

Scene: Saving the Building and Loan

George returns to find a mob standing outside the Building and Loan and he lets the crush of people into the lobby. He then talks with Uncle Billy (Thomas Mitchell) who tells George that he has handed over most of their cash to the bank and closed for the day to prevent a riot. George pleads with the throng while he tries to explain how the banking system works. The Building and Loan doesn't keep all the depositors money on hand; it lends the money out into the community as an investment. George's education lesson does little to quell the tensions and the depositors are about the sell their shares to Mr. Potter until George's wife, Mary (Donna Reed), offers up $2000 in honeymoon money to satisfy the depositors. As a result, the Building and Loan makes it through the remainder of the day and the crisis is averted.

What to Watch for and Ask Yourself

- How do banks make a profit?
- What is different about the financial industry today, which did not exist in the 1940s, that makes a bank run less likely?
- In the scene, George implores the people not to sell their shares for 50 cents on the dollar. "Don't you see what's happening? Potter isn't selling. Potter's buying!" Comment.
- How do banks create money in the economy?

Concepts or Examples Please check the concepts you identify in the clip.

- ☐ Balance sheet
- ☐ Assets
- ☐ Liabilities
- ☐ Fractional reserve banking
- ☐ Money creation
- ☐ Reserves
- ☐ Bank panics
- ☐ Federal Deposit Insurance Corporation (FDIC)

Analysis Write a short analysis using the concepts you checked.

____

____

____

____

____

Personal Reactions

____

____

____

____

____

Traffic

Color, 2000
Clip Running Time: 1 minute, 36 seconds
Rating: R
Director: Steven Soderbergh
Distributor: Universal Studios Home Video

Winner of four Academy Awards, *Traffic* is a critically acclaimed look at the drug industry in the United States and Mexico. The film traces America's War on Drugs through four separate but connected stories. According to the film, efforts to contain drug use function as a price support mechanism for the drug industry. Despite hard work, drugs enter the country, and it is possible for those who wish to get drugs to find them easily. Or as one of the teenagers in the film says, "for someone my age, it's a lot easier to get drugs than it is to get alcohol." *Traffic* relentlessly hammers home the point that the War on Drugs, as currently constituted, cannot be won. Steven Soderbergh, who won Best Director for his efforts, uses a fast-paced documentary style approach to keep the audience engaged. Benico Del Toro won Best Supporting Actor for his role as Javier Rodriguez, a Mexican police officer, intent on fighting crime. *Traffic* also stars Michael Douglas, Catherine Zeta-Jones, Don Cheadle, and Dennis Quaid.

Scene: The Mexican Border

Eduardo Ruiz (Miguel Ferrer) describes his operation to two officers, Montel Gordon (Don Cheadle) and Ray Castro (Luis Guzman). Ruiz explains that drug enforcement in Mexico is an entrepreneurial activity. He bribes customs officials, hires drivers, and throws a lot of product at the problem. Using regression analysis, he is able to identify weaknesses in the border inspection process to help identify practices that will help avoid detection.

What to Watch for and Ask Yourself

- If Ruiz is correct, then NAFTA will expand the flow of drugs into the United States. How does NAFTA impact the circular flow?
- If transactions in the underground economy increase, what happens to GDP?
- Besides the underground economy, what other factors are shortcomings in measuring GDP?
- Drug dealers do not pay income tax. However, that doesn't mean that participants in the underground economy can't be forced to pay taxes like the rest of us. Can you think of a taxation mechanism that would be more effective in capturing revenue from illegal transactions?

Concepts or Examples Please check the concepts you identify in the clip.

- ☐ Gross Domestic Product (GDP)
- ☐ Disposable income
- ☐ Noneconomic sources of well-being
- ☐ North American Free Trade Agreement (NAFTA)
- ☐ Underground Economy
- ☐ Circular flow
- ☐ Income approach
- ☐ Expenditures approach

Analysis Write a short analysis using the concepts you checked.

____

____

____

____

____

Personal Reactions

____

____

____

____

____

Babe

Color, 1995
Clip Running Time: 2 minutes, 12 seconds
Rating: G
Director: Chris Noonan
Distributor: Universal Studios Home Video

Set in Australia, Babe (Christine Cavanaugh, voice) is an orphan pig who escapes the frying pan after being won in a raffle at a country fair by Farmer Hoggett (James Cromwell). The movie is told in a series of chapters and narrated by Roscoe Lee Browne. On the farm, Babe is adopted by Fly (Miriam Margoyles, voice), a border collie who teaches him the farm hierarchy and encourages Babe to find his place. Babe is unaware that the farmer's wife plans on using him for Christmas dinner, but he finds a way to save himself by proving himself more valuable as a sheep herder!

Scene: The Mechanical Roster

Ferdinand the Duck (Danny Mann, voice) devises a plan to avoid being slaughtered by taking over the rooster's crowing responsibilities. However, his plan backfires when Farmer Hoggett's wife purchases an alarm clock. Unless Ferdinand can find a way to eliminate the alarm clock, he is cooked. So Ferdinand solicits help from Babe to enter the house and steal the "mechanical rooster."

What to Watch for and Ask Yourself

- Describe how specialization works on the farm.
- Why is being "indispensable" so important to Ferdinand?
- How does the introduction of the "mechanical rooster" alter the specialization of the farm animals?
- Ferdinand is basing his hopes on replacing the rooster. Does Ferdinand have a comparative advantage in performing the rooster's duties?

Name ____________________

Concepts or Examples Please check the concepts you identify in the clip.

- ☐ Opportunity cost
- ☐ Choice
- ☐ Absolute advantage
- ☐ Comparative advantage
- ☐ Specialization
- ☐ Division of labor
- ☐ Trade
- ☐ Mutual gains from exchange

Analysis Write a short analysis using the concepts you checked.

Personal Reactions

About a Boy

Color, 2002
Clip Running Time: 50 seconds
Rating: PG-13
Director: Chris Weitz and Paul Weitz
Distributor: Universal Studios Home Video

Based on Nick Hornby's popular British novel, *About a Boy* stars Hugh Grant as Will Lightman, a thirty-something Londoner whose purpose in life is to hook up with women. When he is asked about what he does for a living, he responds that he does nothing! His late father wrote a Christmas jingle called "Santa's Super Sleigh" and he lives quite comfortably off of the royalties. As you might expect, Will's flat has every imaginable toy and his shallow existence appears to be going quite well by all appearances. Will is content to take from life, giving back nothing tangible in return. However, when his friends begin to marry off he realizes that he might have to change his game plan in order to continue to meet women. Will decides to pass himself off as a single father, begins attending single parents meetings, invents an imaginary son, and he starts to meet a string of single moms. However, Will's carefree lifestyle is challenged when he meets Marcus (Nicholas Hoult), a quirky twelve-year-old boy who is teased at school. Will teaches Marcus how to be cool and Marcus teaches Will to grow up.

Scene: Island Living

In this scene we see Will going through his daily routine. He is so adept at time management that he has invented a term for his existence, "island living." He is his own activities director, and he has divided the day into 30-minute increments to better organize his life.

What to Watch for and Ask Yourself

- Do you think Will's wealth has anything to do with his lifestyle and choice not to work?
- Will is very good at dividing his life into "units of time." However, he never mentions any activities that require a substantial investment. What kinds of decisions seem to be missing from Will's life?
- What does Will's consumption function look like?
- Will wonders, "if I'd really have time for a job, how do people cram them in?" How does Will's decision to live in the moment impact his chances at a more rewarding life?

Concepts or Examples Please check the concepts you identify in the clip.

- ☐ Short run
- ☐ Long run
- ☐ Consumption function
- ☐ Saving
- ☐ Wealth effect
- ☐ Planned investment
- ☐ Break-even income
- ☐ Utility

Analysis Write a short analysis using the concepts you checked.

____

____

____

____

____

Personal Reactions

____

____

____

____

____

Monty Python's The Meaning of Life

Color, 1983
Clip Running Time: 1 minute, 56 seconds
Rating: R
Director: Terry Gilliam and Terry Jones
Distributor: Universal Studios Home Video

The film is split into seven chapters that sketch the succession of life from birth to death using comedy and music. Along the way the Monty Python crew pokes fun irreverently at the progression from young to old. *The Meaning of Life* is the type of film you either love or hate. The film is a black comedy with two especially memorable scenes. I won't ruin the fun for you. If you enjoy watching the scene in this chapter, I'd recommend that you see the entire film or one of the other notable films by the Monty Python gang, *Life of Brian* (a satire of the birth of Christ) or *Monty Python and the Holy Grail* (based on the legend of King Arthur).

Scene: Just One More Bite

The Mr. Creosote sketch is among the most famous scenes from the film. Mr. Creosote (Terry Jones) is the world's fattest man; he eats all the time and must vomit to ingest more food. This is one of the grossest scenes in film history, and it is easy to see why the sketch did not appear on the popular BBC comedy *Monty Python.* If that wasn't bad enough, Mr. Creosote explodes after eating everything on the menu in a fancy French restaurant. He is egged on to his demise by the maitre d' (John Cleese) who tempts him to eat one last bite: a single wafer thin mint.

What to Watch for and Ask Yourself

- Most fancy restaurants serve small portions so that you can taste many different flavors and textures. Is it surprising to find Mr. Creosote at a fancy restaurant when he could have gone to a buffet and enjoyed all he wanted at a flat price?
- Mr. Creosote has to vomit in order to eat more. Do you think that he is optimizing his consumption choices?
- Ordinarily gains from exchange make both sides better off. However, in this case, Mr. Creosote looks decidedly worse off after consuming his meal. Why does this transaction violate our usual assumptions about gains from trade?
- When should someone stop consuming food? Why does this point occur earlier for most people than Mr. Creosote?

Concepts or Examples Please check the concepts you identify in the clip.

- ☐ Utility
- ☐ Total utility
- ☐ Average utility
- ☐ Marginal utility
- ☐ Diminishing marginal utility
- ☐ Optimizing consumption choices
- ☐ Substitution effect
- ☐ Consumer surplus

Analysis Write a short analysis using the concepts you checked.

____

____

____

____

____

Personal Reactions

____

____

____

____

____

Being John Malkovich

Color, 1999
Clip Running Time: 3 minutes, 54 seconds
Rating: R
Director: Spike Jonze
Distributor: Universal Studios Home Video

Craig Schwartz (John Cusack) is a puppeteer who is struggling to make a living so he answers an unusual help-wanted ad and the adventure ensues. *Being John Malkovich* (nominated for three Academy Awards) is a wonderful film with twists and unexpected developments at every corner. One day Craig discovers a small opening behind a filing cabinet. He crawls through it and enters a portal into the mind of actor John Malkovich (played by himself). There he stays, experiencing exactly what Malkovich sees for 15 minutes, until he is unceremoniously dumped out of the sky next to the New Jersey Turnpike. When Craig mentions his discovery to Maxine (Catherine Keener), she helps him make it commercial business by selling trips inside Malkovich for $200 a pop. However, it doesn't take long for the new enterprise to go haywire.

Scene: The Portal

Malkovich discovers that Craig and Maxine are profiting from the portal and he goes to investigate. Upon arriving at the business he encounters a long line of people waiting to enter the portal. Not willing to wait, Malkovich cuts through the line and confronts Craig and Maxine. After they describe the business to Malkovich, he decides that he wants to try the portal out himself.

What to Watch for and Ask Yourself

- Does $200 for 15 minutes of fantasy strike you as a high or low price?
- Is the business competitive, oligopolistic, or a monopoly?
- What makes the service that is being sold unique?
- In the scene, there is a long queue of people waiting to experience the life of Malkovich. This is a good indicator of excess demand. How do markets typically react when there is more demand than supply?

Concepts or Examples Please check the concepts you identify in the clip.

- ☐ Monopoly
- ☐ Excess demand
- ☐ Competition
- ☐ Availability of substitutes
- ☐ Unique goods
- ☐ Exploitation
- ☐ Sources of profit
- ☐ Pricing power

Analysis Write a short analysis using the concepts you checked.

____

____

____

____

____

Personal Reactions

____

____

____

____

____

Intolerable Cruelty

Color, 2003
Clip Running Time: 2 minutes, 39 seconds
Rating: PG-13
Director: Joel Coen
Distributor: Universal Studios Home Video

Miles Massey (George Clooney) is a divorce attorney who is widely respected among his peers for creating an unbreakable prenuptial agreement. Miles wins a case against Marilyn Rexroth (Catherine Zeta-Jones) after exposing her plan to marry for money. Marilyn, it turns out, has a history of wealthy husbands. So it is surprising when Marilyn wants to hire Miles to draft an iron-clad prenuptial agreement for her next marriage. Miles suspects that Marilyn is only marrying millionaire oil man Howard D. Doyle (Billy Bob Thornton) for his money, but he can't figure out why she would want a prenuptial agreement.

Scene: A Prenuptial Agreement

Marilyn walks into Miles' office and introduces her fiancé, Howard Doyle. As a gesture of good faith Marilyn indicates to Miles that she would like to sign off on a prenuptial agreement. Miles then advises Marilyn and Howard that the prenuptial agreement will mean that the possessions that they bring into the marriage will return to them if the marriage is dissolved. Also, any earnings made during the marriage will accrue to the person who earned them.

What to Watch for and Ask Yourself

- Why does asymmetric information create a market for prenuptial agreements?
- A prenuptial agreement specifies how assets are to be divided in the event of a divorce. Is this an example of a positive, negative, or zero-sum game?
- Is a prenuptial agreement designed to eliminate adverse selection or moral hazard?
- What are some other examples of asymmetric information and how does the market try to correct for information failures?

Concepts or Examples Please check the concepts you identify in the clip.

- ☐ Asymmetric information
- ☐ Moral hazard
- ☐ Adverse selection
- ☐ Cooperative games
- ☐ Positive-sum games
- ☐ Zero-sum games
- ☐ Negative-sum games
- ☐ Risk tolerance
- ☐ Financial intermediation

Analysis Write a short analysis using the concepts you checked.

____

____

____

____

____

Personal Reactions

____

____

____

____

____

Along Came Polly

Color, 2004
Clip Running Time: 2 minutes, 25 seconds
Rating: PG-13
Director: John Hamburg
Distributor: Universal Studios Home Video

Rueben Feffer (played by Ben Stiller) is an insurance analyst who decides to take a chance when he meets an old high school classmate, Polly Prince (Jennifer Aniston). Rueben is on the rebound after learning that his wife cheated on him during their honeymoon. Polly is a free spirit who cannot commit to a long-term relationship. The pair make an odd couple. Rueben is constantly worried about risk while Polly enjoys life, with little care for playing it safe. John Hamburg also directed Ben Stiller in *Meet the Parents* and *Zoolander.*

Scene: Liberating Reuben

Rueben is fluffing throw pillows and placing them on his bed like he does everyday when Polly asks why he spends so much time placing and replacing the pillows. Rueben has never really thought about all the time it takes to keep the bed looking nice. Polly decides to liberate Rueben by taking a knife to one of the throw pillows to prove her point. Later we see that Polly has lost her keys, so she uses a key finder that Reuben has given her to locate them inside her refrigerator.

What to Watch for and Ask Yourself

- Rueben's place is spotless. Can you be too neat?
- What is the cost of trying to attain perfection (in our looks, the environment, safety)?
- Is perfection a realistic economic goal?
- Polly's place is cluttered and unkempt. Do you think it is okay to be a little bit messy?
- Efficient behavior demands that the marginal benefits be equal to the marginal costs. Therefore, optimal behavior is influenced by the trade-off between the added benefits and the additional costs of an activity.

Concepts or Examples Please check the concepts you identify in the clip.

- ☐ Production efficiency
- ☐ Allocative efficiency
- ☐ Optimal behavior
- ☐ Equal marginal principle
- ☐ Consumer optimum
- ☐ Marginal cost
- ☐ Marginal benefit
- ☐ The law of increasing opportunity costs

Analysis Write a short analysis using the concepts you checked.

____

____

____

____

____

Personal Reactions

____

____

____

____

____

In the Name of the Father

Color, 1993
Clip Running Time: 2 minutes, 54 seconds
Rating: R
Director: Jim Sheridan
Distributor: Universal Studios Home Video

In the aftermath of September 11, 2001 and the ongoing War on Terror, *In the Name of the Father*, conveys a powerful message: Wrongful imprisonment and a lack of checks and balances in the legal system are a threat to personal freedom. In the early 1970s, the British government passed the Prevention of Terrorism Act, which allowed the arrest of any individual on the thinnest of suspicions. A young homeless Irishman, Gerry Conlon (Daniel Day-Lewis, 1989 best actor for *My Left Foot*), and three of his companions were charged by the British police with being IRA bombers for destroying a pub in Guildford, England. The Guilford Four, as they were later called, were falsely convicted and sentenced to life. Moreover, evidence that would have acquitted the Guilford Four was suppressed. However, through the efforts of a stubborn attorney, Gareth Peirce (Emma Thompson), the verdict is ultimately overturned and the Four were eventually released in 1989. The film is based on Conlon's autobiography, *Proved Innocent*.

Scene: Forcing a Confession

Gerry Conlon is the prime suspect in the Guilford bombings. After a brutal and sadistic interrogation by the police, Conlon breaks and agrees to sign a confession. In the scene we see the police interrogate Gerry and his friends, Carole Richardson (Beatie Edney), Paddy Armstrong (Mark Sheppard), and Paul Hill (John Lynch). The police interrogate each of the suspects separately and try to break them both physically and psychologically. This creates a prisoner's dilemma where the suspects do not know what the others are saying about them to the police.

What to Watch for and Ask Yourself

- Would the police have been able to force confessions as easily if the suspects had been interrogated as a group?
- In what ways is the prisoner's dilemma an information problem?
- Did the suspects follow the dominant strategy in the interrogation process?
- Part of the prisoner's dilemma is that they are placed in a non-cooperative setting. This makes collusive behavior difficult. What conditions foster collusion?

Concepts or Examples Please check the concepts you identify in the clip.

- ☐ Cooperative game
- ☐ Game theory
- ☐ Zero-sum game
- ☐ Prisoner's dilemma
- ☐ Collusion
- ☐ Bargaining power
- ☐ Dominant strategy
- ☐ Opportunistic behavior

Analysis Write a short analysis using the concepts you checked.

____

____

____

____

____

Personal Reactions

____

____

____

____

____

Bread and Roses

Color, 2000
Clip Running Time: 2 minutes, 15 seconds
Rating: R
Director: Ken Loach
Distributor: 16 Film Limited

The culture of greed in Southern California is exposed through the lives of Maya (Pilar Padilla) and Rosa (Elpidia Carillo), two illegal emigrants from Mexico, who end up working as janitors in Los Angeles. Enter Sam Shapiro (Adrien Brody, 2002 Best Actor for *The Pianist*), a union organizer who tries to mount a "justice for janitors" campaign to organize the workers. Maya is a willing follower, but Rosa is reluctant since she has her ailing husband to consider. In addition, Maya and Rosa are plagued by their boss (George Lopez) who is a foul-mouthed and quick-tempered supervisor with the power to fire workers on a whim. So the threat of reprisals for union activity is a very real possibility for the workers.

Scene: Strike!!!

Sam has organized a demonstration in the building lobby as a show of worker solidarity. Non-unionized janitors are being paid $5.75 an hour, have no benefits, and are laid off indiscriminately. This is in stark contrast with unionized janitors who are paid over $8 an hour with medical coverage. By forming a union, the janitors hope to achieve better pay and benefits.

What to Watch for and Ask Yourself

- What do "bread" and "roses" stand for in the scene?
- According to the scene, when do you get "roses"?
- What does Sam mean when he advocates a "labor union that is strong enough to compete against the companies that are controlling our lives"?
- Sam mentions the historical roots of the "bread and roses" slogan. In 1912, the Industrial Workers of the World (IWW) organized a strike of textile workers. Most of the workers were immigrants so the parallel with the "justice for janitors" campaign makes sense, not only as a union effort but also an attempt to bring justice to a group that historically has struggled to achieve fair wages.

Concepts or Examples Please check the concepts you identify in the clip.

- ☐ Strike
- ☐ Minimum wage
- ☐ Collective Bargaining
- ☐ Real wages
- ☐ Compensating differentials
- ☐ Labor supply
- ☐ Monopsony
- ☐ Elasticity of demand for inputs

Analysis Write a short analysis using the concepts you checked.

____

____

____

____

____

Personal Reactions

____

____

____

____

____

Parenthood

Color, 1989
Clip Running Time: 2 minutes, 8 seconds
Rating: PG-13
Director: Ron Howard
Distributor: Universal Studios Home Video

Steve Martin stars as Gil Buckman in a role that transitioned Martin from a slapstick actor to one capable of showing a broad range of emotions. Gil is a father who is determined to raise his kids the right way. As the film unfolds, we begin to realize that Gil's desire to raise his kids is a reaction to the neglect that he experienced as a child. So in order to be the best possible dad, Gil sacrifices his career ambitions. As a result, his family is tight financially. Steve Martin does a wonderful job of portraying the conflicting emotions of parenthood. He wants to be a good father and a good provider, but both are not possible simultaneously. Ron Howard directs and he does a masterful job of blending together four generations of the family in order to explore parent-child relationships on many different levels.

Scene: A Broken Promise

When Gil learns that his rival, Phil Richards, has passed him over for a position as a partner in the firm, Gil decides to vent his frustration to his boss, David Brodsky (Dennis Dugan) and plead his case. David responds that the decision is not etched in stone but that Phil was selected because he works longer, schmoozes better, and brings in more business. Gil counters that he is the backbone of the company and that he might have to look for another job, to which David replies that he doesn't consider Gil's threat to leave to be a credible threat. David believes that Gil is trapped because he is not committed enough and that he is not willing to take a financial hit in order to start over somewhere else. This exchange raises a number of important issues about the nature of the labor market and how much workers are compensated for their services.

What to Watch for and Ask Yourself

- Who has the highest marginal revenue product, Phil or Gil?
- How much is Gil's experience worth in the scene?
- What would give Gil additional leverage in asking for a raise or promotion?
- Discuss the factors that go into wage determination.

Concepts or Examples Please check the concepts you identify in the clip.

- ☐ Marginal physical product
- ☐ Derived demand
- ☐ Determinants of demand elasticity for inputs
- ☐ Wage determination
- ☐ Marginal revenue product
- ☐ Substitutability of inputs
- ☐ Comparative advantage
- ☐ Experience

Analysis Write a short analysis using the concepts you checked.

Personal Reactions

Erin Brockovich

Color, 2000
Clip Running Time: 2 minutes, 3 seconds
Rating: R
Director: Steven Soderbergh
Distributor: Universal Studios Home Video

Julia Roberts won Best Actress for her portrayal of Erin Brockovich. The film is based on the true story of Brockovich, a single mother with no formal education, few prospects for employment, and down on her luck. Eventually, she gets a job as a file clerk for a small law firm. There she begins to investigate (on her own) illnesses in a small town caused by the illegal dumping of deadly toxic waste by Pacific Gas & Electric. Spurred on by her efforts at uncovering the truth, the law firm she works for becomes involved in one of the largest class action lawsuits in history, resulting in a settlement of over $300 million! *Erin Brockovich* is one of only a handful of films that have tackled environmental issues in a meaningful fashion; the others are *Silkwood* and *A Civil Action*.

Scene: An Offer from PG&E

Ed Masry (Albert Finney) and Erin Brockovich meet with a representative of Pacific Gas & Electric to discuss a settlement offer from the firm. The company offer compensates the claimants for the value of the land they own, but no compensation is offered for medical damages as a result of toxic poisoning. Erin Brockovich believes that many of the health problems in the affected community can be traced to the introduction of the poison into the groundwater. However, PG&E counters that the condition of the residents is a result of lifestyle choices, heredity, and bad luck.

What to Watch for and Ask Yourself

- When a third party is adversely affected by the actions of others this is referred to as what type of externality?
- PG&E decided to dump the toxic poison, rather than clean it up. Describe how this decision lowered the firm's private costs but raised the social cost to society.
- If society banned the production of toxic poisons would we be better off?
- How does the issue of environmental pollution relate to the enforcement or ownership of property rights?

Concepts or Examples Please check the concepts you identify in the clip.

- ☐ Externalities
- ☐ Market failure
- ☐ Internalization of external costs
- ☐ Social costs
- ☐ Spillover costs
- ☐ Optimal quantity of pollution
- ☐ Property rights
- ☐ Private costs

Analysis Write a short analysis using the concepts you checked.

____

____

____

____

____

Personal Reactions

____

____

____

____

____

The River

Color, 1984
Clip Running Time: 3 minutes, 21 seconds
Rating: P-13
Director: Mark Rydell
Distributor: Universal Studios Home Video

Farming as a way of life was under pressure in the early 1980s when interest rates skyrocketed and the economy was in recession. Tom (Mel Gibson) and Mae (Sissy Spacek) Garvey own a small farm in Tennessee where they must battle storms and repossession to make ends meet. The Garvey's struggles are magnified by the fact that their property is in a flood plain, and a local developer wants to dam up the river and flood the fields to build a hydroelectric project. Eventually, things get so bad financially that Tom is forced to take a job as a scab at a local foundry. The contrast between the natural beauty of the farm and the gritty inferno at the steel mill give the viewer a deep sense of appreciation for the why the Garvey's fight so hard to preserve their way of life.

Scene: A Bankruptcy Auction

Tom and Mae Garvey attend an auction for Dan Gaumer (Jim Antonio). When the auctioneer invites the assembled crowd to gather a number of the farmers begin to chant, "No sale! No sale!" to prevent the sale of the Gaumer farm. However, it is Dan Gaumer who eventually quiets the crowd by saying that the auction is not for the farm; it is for the personal property, machinery, and livestock that his family still owns. The bank foreclosed on the farm and the family needs to sell their property to relocate elsewhere.

What to Watch for and Ask Yourself

- Do you think that the Gaumer's will get fair value for the personal property that they are selling? If not, do you think they will receive more or less than what they paid?
- Why do you think that the bank foreclosed on the property?
- Could the Gaumer's have done anything to prevent the foreclosure?
- The crowd does not want the Gaumer family to lose their farm, yet there are a large number of people attending the auction. Is it possible to reconcile this inconsistency?
- How does the elimination of one farmer affect those who remain in business? Evaluate using the perfectly competitive theory of the firm.

Concepts or Examples Please check the concepts you identify in the clip.

- ☐ English auction
- ☐ Value
- ☐ Supply and demand
- ☐ Shortages and surpluses
- ☐ Profits
- ☐ Property Rights
- ☐ Rationing mechanisms
- ☐ Scarcity

Analysis Write a short analysis using the concepts you checked.

Personal Reactions

Bibliography

American Film Institute. 1998. Top 100 Films of the 20th Century. (http://www.filmsite.org/afi100films.html) (http://www.afi.com/tvevents/100years/movies.aspx)

Becker, W. E., and Watts, M. 1998. *Teaching economics to undergraduates: Alternatives to chalk and talk*. Cheltenham: Edward Elgar.

Becker, W. E., and Watts, M. 2001. Teaching methods in U.S. undergraduate courses. *Journal of Economic Education* 32 (Summer): 269–280.

Becker, W. E. 2003. How to make economics the sexy social science. *Southern Economic Journal* 70 (Summer): 195—198.

Bleiler, D., ed. 2004. *TLA Video & DVD Guide: The Discerning Film Lover's Guide*. New York: St. Martin's.

Ebert, R., ed. 2003. *Roger Ebert's Movie Yearbook 2004*. Kansas City: Andrews McMeel Publishing.

Formaini, R. 2001. Free markets on film: Hollywood and capitalism. *Journal of Private Enterprise* 16 (Spring): 122–129.

Internet Movie Database. (http://www.imdb.com/)

Kish-Goodling, D. M. 1998. Using *The Merchant of Venice* in teaching monetary economics. *Journal of Economic Education* 29 (Fall): 330–339.

Leet, D. and Houser, S. 2003. Economics goes to Hollywood: Using classic films to create an undergraduate economics course. *Journal of Economic Education* 34 (Fall): 326–332.

Maltin, L., ed. 2003. *Leonard Maltin's 2004 Movie & Video Guide*. New York: Penguin.

Mulder, C. 2002. Current Labor Issues in Film. (http://fandm.edu/departments/tdf/FilmPages/Courses/Other-02/syllabus-2002.htm)

Ribstein, L. 2004. Busfilm. (http://busmovie.typepad.com/busfilm/)

Tinari, F. D., and Khandke, K. 2000. From rhythm and blues to Broadway: Using music to teach economics. *Journal of Economic Education* 31 (Summer): 253–270.

Shor, M. 2004. Resources for learning and teaching strategy for business and life. (http://www.gametheory.net)

Watts, M. 1998. Using literature and drama in economics courses. In *Teaching economics to undergraduates: Alternatives to chalk and talk*. 185–207. Cheltenham: Edward Elgar.

Index